HOT AND COLD ANIMALS

DESERT HARE OR ARCTIC HARE

BY ERIC GERON

Children's Press
An imprint of Scholastic Inc.

D1378148

A special thank you to the team at the Cincinnati Zoo & Botanical Garden for their expert consultation.

Library of Congress Cataloging-in-Publication Data
Names: Geron, Eric, author.
Title: Hot and cold animals. Desert hare or Arctic hare / by Eric Geron.
Other titles: Desert hare or Arctic hare
Description: First edition. | New York : Children's Press, an imprint of Scholastic Inc., 2022. | Series: Hot and cold animals | Includes index. | Audience: Ages 5–7. | Audience: Grades K–1. | Summary: "NEW series. Nonfiction, full-color photos and short blocks of text to entertain and explain and how some animals with the same name can survive in very different environments"—Provided by publisher.
Identifiers: LCCN 2021044805 (print) | LCCN 2021044806 (ebook) | ISBN 9781338799453 (library binding) | ISBN 9781338799460 (paperback) | ISBN 9781338799477 (ebk)
Subjects: LCSH: Hares—Juvenile literature. | Arctic hare—Juvenile literature. | Habitat (Ecology)—Juvenile literature. | BISAC: JUVENILE NONFICTION / Animals / Rabbits | JUVENILE NONFICTION / Animals / General
Classification: LCC QL737.L32 G47 2022 (print) | LCC QL737.L32 (ebook) | DDC 599.32/8—dc23
LC record available at https://lccn.loc.gov/2021044805
LC ebook record available at https://lccn.loc.gov/2021044806

Copyright © 2022 by Scholastic Inc.

All rights reserved. Published by Children's Press, an imprint of Scholastic Inc., *Publishers since 1920.* SCHOLASTIC, CHILDREN'S PRESS, and associated logos are trademarks and/or registered trademarks of Scholastic Inc.

The publisher does not have any control over and does not assume any responsibility for author or third-party websites or their content.

No part of this publication may be reproduced, stored in a retrieval system, or transmitted in any form or by any means, electronic, mechanical, photocopying, recording, or otherwise, without written permission of the publisher. For information regarding permission write to Scholastic Inc., Attention: Permissions Department, 557 Broadway, New York, NY 10012.

10 9 8 7 6 5 4 3 2 1   22 23 24 25 26

Printed in the U.S.A. 113
First edition, 2022

Book design by Kay Petronio

Photos ©: cover right and throughout: All Canada Photos/Alamy Images; 4: Tim Fitzharris/Minden Pictures; 5: Robert Postma/Design Pics/age fotostock; 6–7: Supercaliphotolistic/Getty Images; 8–9: Matthias Breiter/Minden Pictures; 11: Darren Keast/500px/Getty Images; 14–15: Matthias Breiter/Minden Pictures; 16 bottom: Jim McMahon/Mapman ©; 17 left: Michael S. Nolan/age fotostock; 17 right: Jim McMahon/Mapman ©; 18–19: Bryn Sharp/500px/Getty Images; 20–21: Matthias Breiter/Minden Pictures; 22 main: Terry W. Eggers/Getty Images; 23: Adstock/UIG/age fotostock; 24–25: Katharina Notarianni/Dreamstime; 26–27: Jim Brandenburg/Minden Pictures; 29: Jim Brandenburg/Minden Pictures. All other photos © Shutterstock.

DESERT HARE

ARCTIC HARE

CONTENTS

MEET THE HARES

Desert hares and Arctic hares are different in many ways. Desert hares, also called "black-tailed **jackrabbits**," live in the hot, dry deserts in southwestern North America. They like to rest in cool spots and eat plants.

DESERT HARE

Arctic hares live in the icy, cold **tundra** of North America. They like to dig in the snow and huddle together for warmth.

ARCTIC HARE

FACT Claws help the Arctic hare dig cozy spots in the snow to avoid the cold.

WHERE'S THE HARE?

The desert hare has dark gray fur to camouflage against its surroundings.

WARNING TAIL

The desert hare's tail lifts up to warn other hares of trouble!

LONG LEGS

Strong hind legs make the desert hare an excellent runner and jumper.

BIG EARS

A desert hare's long ears help it hear if a threat is coming close.

A desert hare can weigh 3 to 9 pounds (1.4 to 4 kg).

It has dark gray fur, a white belly, and a black-tipped tail.

FACT

Desert hares can jump as high as 10 feet (3 m) in the air!

ARCTIC HARE CLOSE-UP

An Arctic hare can weigh 6 to 15 pounds (3 to 7 kg).

It has pure white fur in the winter so it can blend in to the snow.

WINTER COAT

Thick fur keeps the Arctic hare warm in the cold.

QUICK AS A BUNNY!

Arctic hares are super-speedy runners due to their strong hind legs!

LITTLE EARS

An Arctic hare's short ears are close to its body to help keep it warm.

TEETH TOOLS

Long front teeth are useful for grabbing plants from hard-to-reach places.

PAW PADS

The thick pads on the Arctic hare's paws prevent the hare from slipping on the ice.

FACT

In the summer, the Arctic hare's white fur will **molt** and change to gray!

LIGHT AND HEAVY

Desert hares are brown and lightweight. Arctic hares are white in winter— and twice as heavy as desert hares!

DESERT HARE

In fact, Arctic hares are the heaviest type of hare in North America.

Desert hares have black fur on their ears and tail, while Arctic hares only have black fur on their ears.

ARCTIC HARE

Desert hares are quicker than Arctic hares and can run as fast as 45 miles per hour (72 km per hour)!

The highest temperature ever recorded in southwestern North America was 128°F (53°C)!

FACT

DRY DESERT DAYS

Desert hares live in the dry, dusty deserts of southwestern North America. Although deserts can get very hot during the day, they can also get very cold at night and in the winter. When desert temperatures dip, the desert hare snuggles up in tall grasses to keep out of the chilly air. While it is often bright and warm where they live, desert hares spend their days out of the sun! They rest in shady spots during the day and move around at night once the weather cools down.

ICE, ICE, BUNNY!

In the winter, the tundra of North America is a very cold place. Arctic hares must use their excellent sense of smell to find food. Although Arctic hares typically live alone, sometimes they feed in large **droves**—with as many as 300 hares at a time! Gathering in droves is also a clever way to huddle together for warmth and to keep a lookout for hungry **predators**.

FACT

In the winter, the average Arctic temperature is −40°F (−40C)!

NIGHTTIME ANIMALS

Desert hares and Arctic hares live in very different places. The tundra of North America has freezing winters. The deserts of southwestern North America have sizzling-hot summers.

Arctic Ocean

North America

Atlantic Ocean

Europe

Asia

Pacific Ocean

Africa

South America

Indian Ocean

Pacific Ocean

Australia

Where desert hares live

Southern Ocean

Antarctica

DESERT HARE

Both hares are **nocturnal**. Nocturnal means an animal is active during the night.

ARCTIC HARE

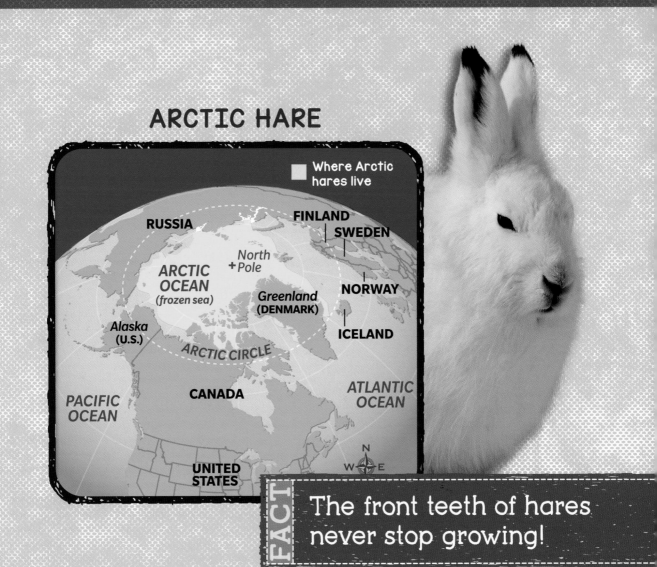

Where Arctic hares live

RUSSIA

FINLAND
SWEDEN

North +Pole

ARCTIC OCEAN (frozen sea)

Greenland (DENMARK)

NORWAY

Alaska (U.S.)

ICELAND

ARCTIC CIRCLE

ATLANTIC OCEAN

PACIFIC OCEAN

CANADA

N
W E
S

UNITED STATES

FACT

The front teeth of hares never stop growing!

Desert hares are one of the fastest animals found on land!

FACT

MUNCH, MUNCH, MUNCH!

Desert hares are **herbivores**. An herbivore is an animal that eats plants. Desert hares mostly eat grasses and plants such as sagebrush and cacti.

NIBBLE, NIBBLE, NIBBLE!

Arctic hares are **omnivores**. An omnivore is an animal that eats plants and meat. In the winter, Arctic hares eat mostly plants, moss, and **lichen**. They use their sense of smell to find food when it is covered by snow. Their claws can dig through the snow. When winters are harsh, they sometimes eat meat and fish. In the summer, they eat plants and berries.

FACT Arctic hares are also called "polar rabbits."

THIRSTY HARES

Desert hares and Arctic hares both need water to survive. There is not much water where both hares live.

DESERT HARE

Coyotes, wolves, foxes, and owls are some of the desert hare's predators.

FACT

Water is hard to find in the dry desert and frozen tundra. Because of this, desert hares must get most of their water from the plants they eat, while Arctic hares chew on ice and snow.

ARCTIC HARE

FACT

Arctic wolves, foxes, lynxes, and owls are some of the Arctic hare's predators.

Baby hares are called **leverets**. They can hop within minutes of being born! **FACT**

SUMMER BABIES

Desert hares usually give birth to two to five babies per **litter**. A litter is a group of babies born at the same time to the same mother. The snuggly bunnies live in a shallow den in the ground. When the baby desert hares are born, they weigh around 3 ounces (85 g). They will go out on their own after one month.

FROSTY BABIES

Arctic hare mothers give birth to two to eight babies per litter. The babies live in a soft little nest in the ground. They can leave the nest when they are two to three weeks old. When they are eight to nine weeks old, they are ready to live on their own.

FACT Arctic hare babies are born with grayish-brown fur.

CUTEST HARES

Although they grow up to be different adult hares, desert hare babies and Arctic hare babies are the same in many ways.

DESERT HARE LEVERET

A male desert hare is called a jack and a female desert hare is called a jill.

They both are born with fur and the ability to see. They both are born in a shallow hole in the ground. Desert hare mothers can give birth four times a year. Arctic hare mothers only give birth once a year, usually in the spring or summer, when food is easier to find.

ARCTIC HARE LEVERETS

YOU DECIDE!

If you had to choose, would you rather be a desert hare or an Arctic hare? If you like hot weather and napping in the shade, you may prefer being a desert hare. If you like cold weather and lots of snow, maybe you would choose to be an Arctic hare!

There are 30 different types of hares.

FACT

GLOSSARY

camouflage (KAM-uh-flahzh) – to disguise something so that it blends in with its surroundings

desert (DEZ-urt) – a dry area where hardly any plants grow because there is so little rain

drove (drove) – a group of hares

herbivore (HUR-buh-vor) – an animal that only eats plants

hind (hinde) – at the back or rear

jackrabbits (JAK-rab-its) – hares found in southwestern North America

leveret (LEV-ur-it) – a baby hare

lichen (LYE-kuhn) – a simple plant that grows on rocks, walls, or trees

litter (LIT-ur) – a number of baby animals that are born at the same time to the same mother

molt (mohlt) – to shed (old feathers or fur)

nocturnal (nahk-TUR-nuhl) – active at night

omnivore (AHM-nuh-vor) – an animal that eats both plants and meat

predator (PRED-uh-tur) – an animal that lives by hunting other animals for food

tundra (TUHN-druh) – a very cold area where there are no trees and the ground is always frozen

INDEX

ABOUT THE AUTHOR

Eric Geron is the author of many books. He lives in Los Angeles, California, with his tiny dog. If he had to choose, he would be a desert hare.